See th

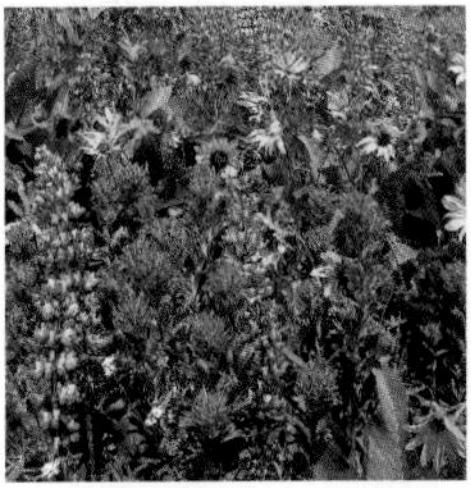
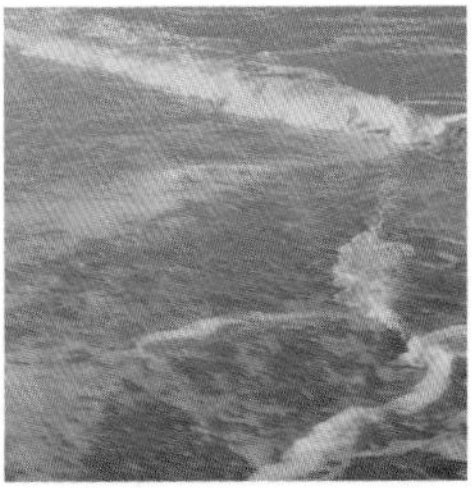
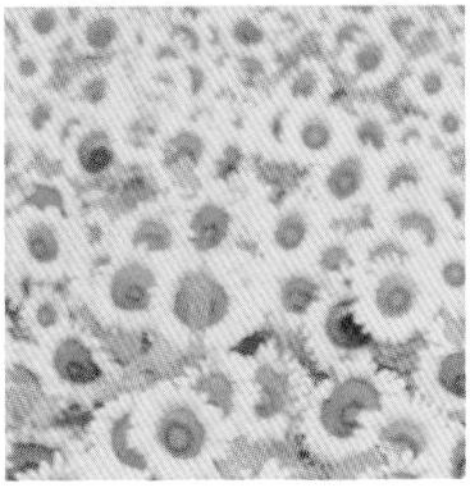
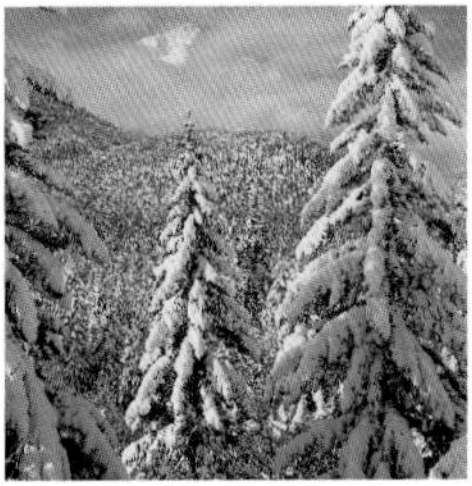

Written by
Rozanne Lanczak Williams

Harcourt

Orlando Boston Dallas Chicago San Diego

www.harcourtschool.com

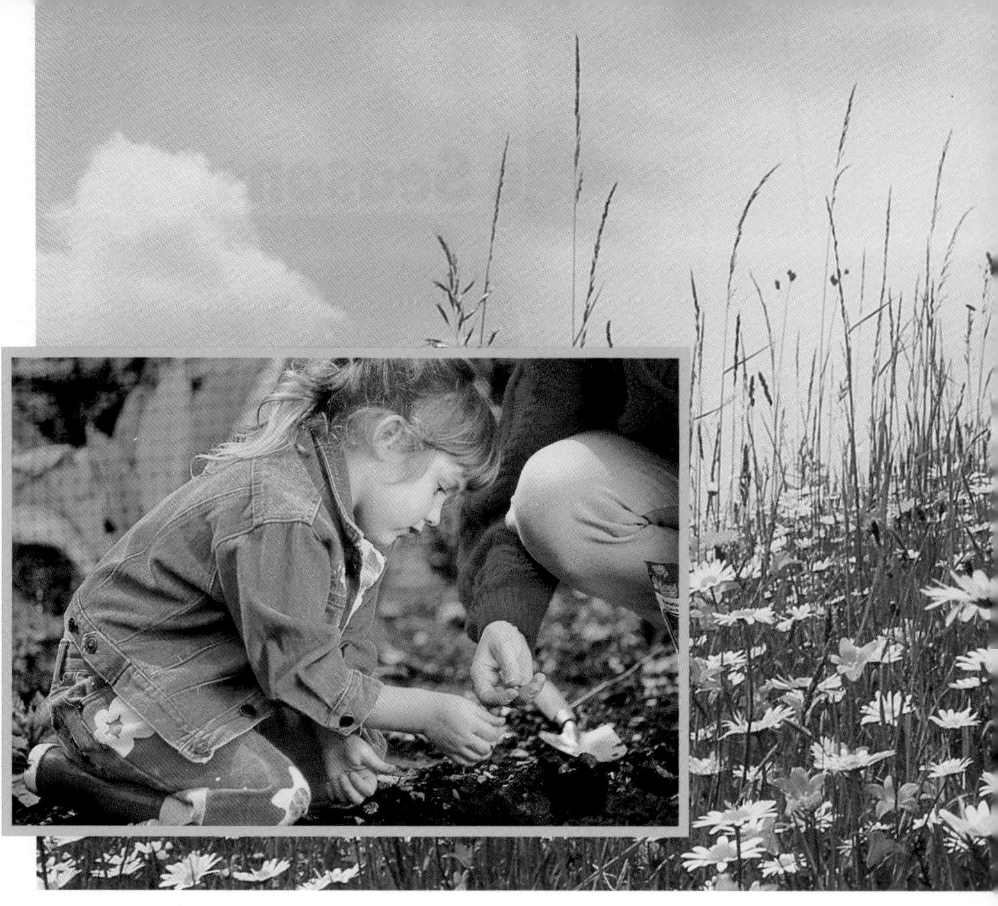

What can you see?

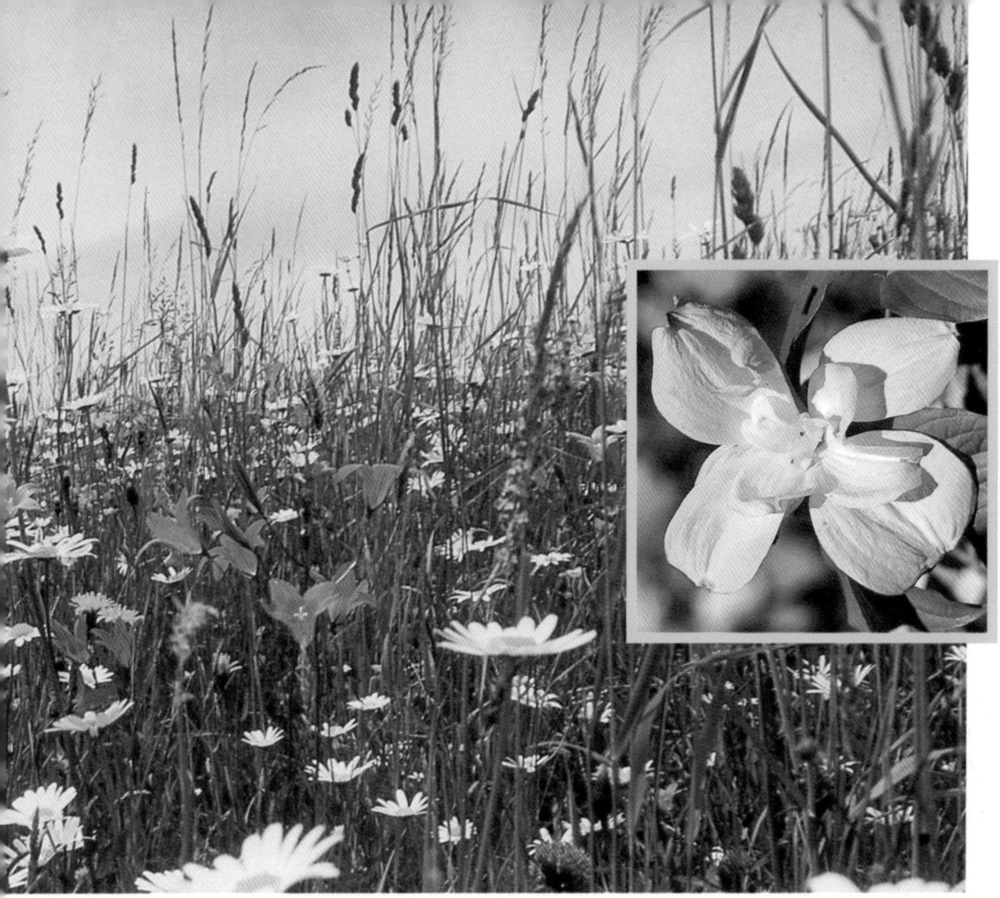

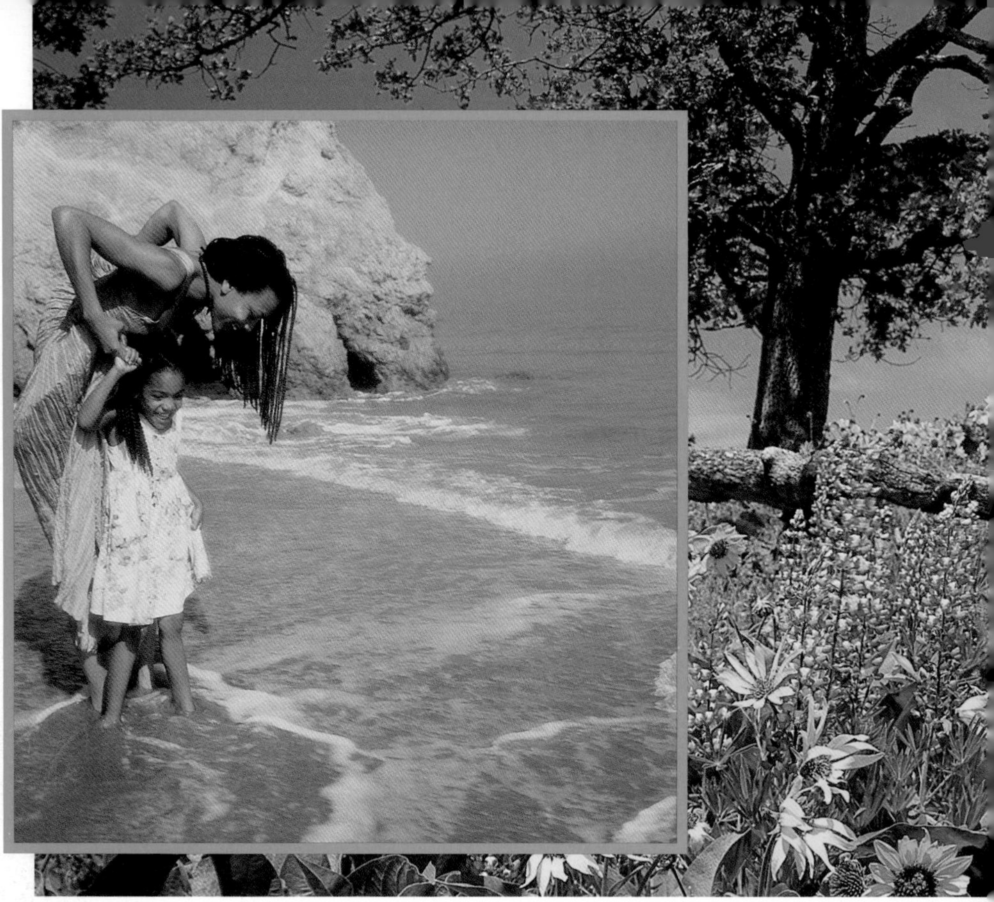

What can you see?

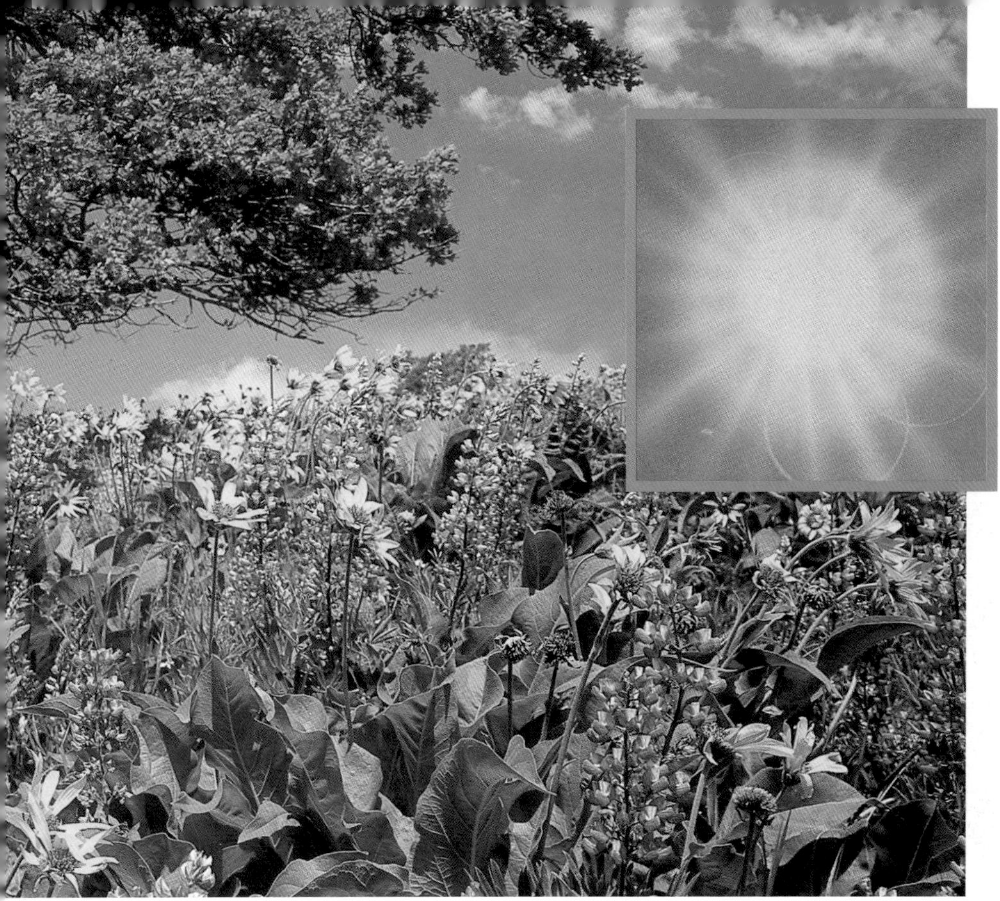

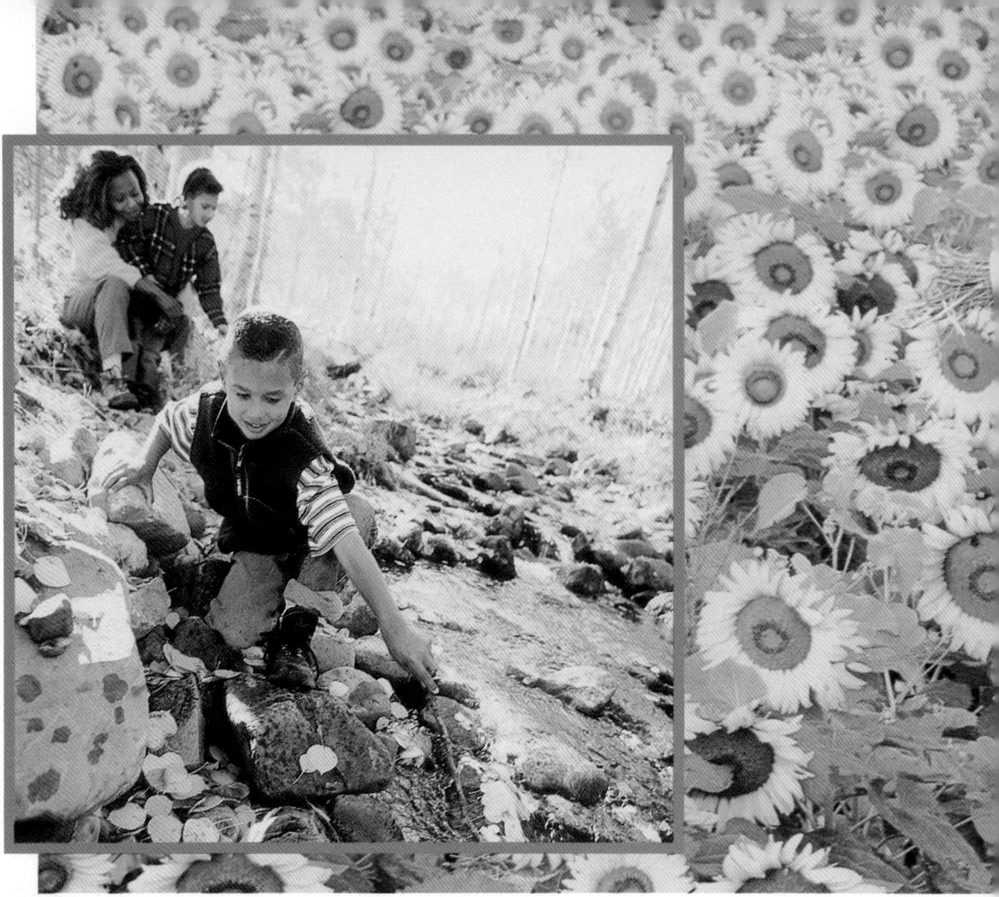

What can you see?

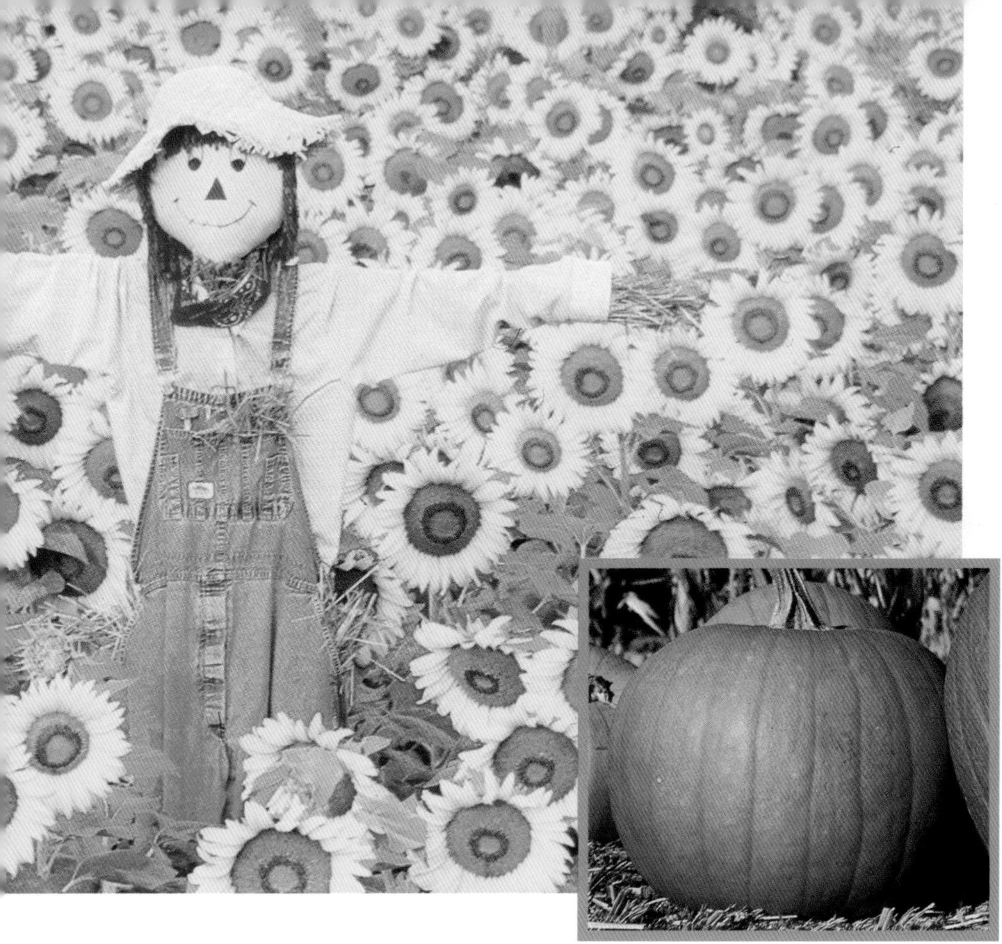

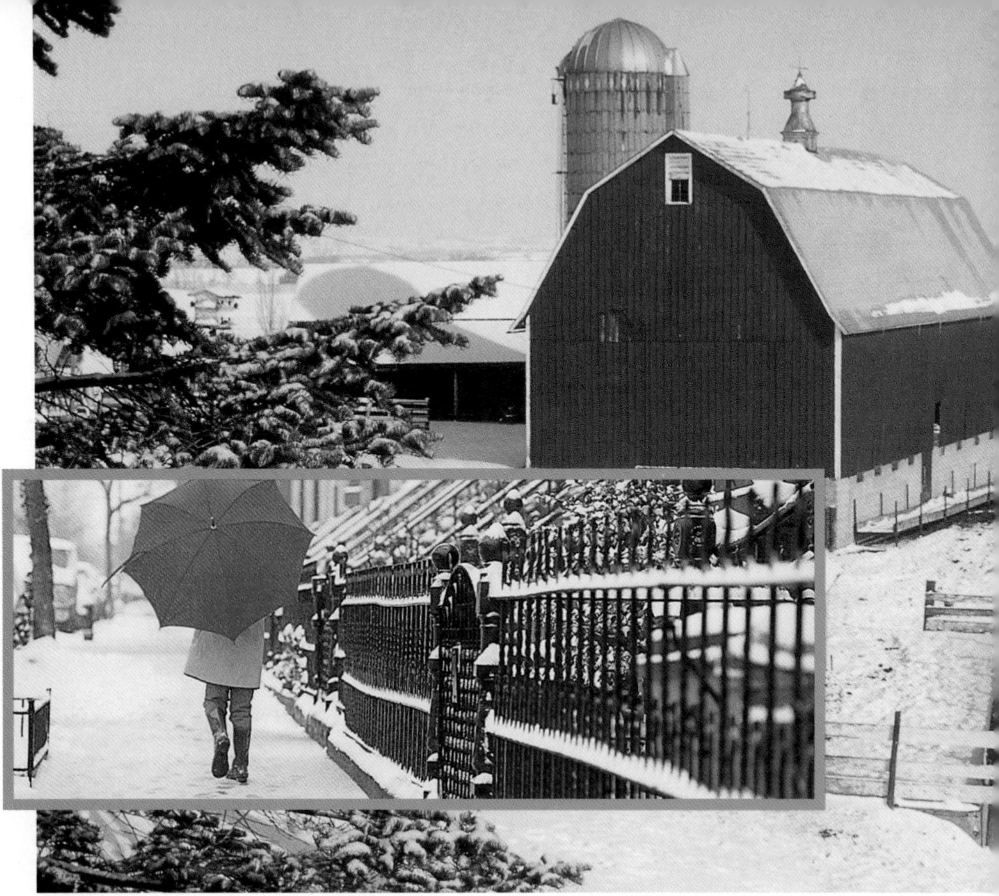

Can you see me?